THE END OF GLOBAL WARMING

WARMING

The Beginning of a New Beginning

Ray Boyd

Global Warming is currently affecting all life on Earth. Without taking control Earth could became a dead planet.

This is dedicated to those on the Global Warming front. It is an uphill battle and they must succeed.

CONTENTS

PROLOGUE

THIS IS BASED ON CURRENT FACTS AND FUTURE FICTION BASED ON FACTS.

The world is in a crisis because of Global Warming when a second crisis arose. An Asteroid is discovered and headed toward earth.

2023. NASA Planetary Defense Coordination Office (PDCO) began tracking Asteroid 2023DW (fact). It was detected and headed toward Earth on a possible collision course to arrive 14 February 2046 (fact). If nothing is done the Earth may not be habitable. Remember the Dinosaurs!

2015. A meeting was held called the Paris Climate Agreement (fact). 144 countries attended and agreed to reduce emissions of greenhouse gases contributing to Global Warming, the annual increase in the temperature of the earth.

Continuously heating the Earth can create a thermal runaway. A runaway greenhouse effect is unlikely to happen naturally. However, it is a possible outcome of human-caused climate change. A runaway greenhouse scenario is when a planet's atmosphere traps so much heat that the planet's temperature rises uncontrollably, on its own. This positive feedback loop can lead to a rapid and irreversible increase in the planet's temperature. The consequences of a runaway greenhouse effect would be catastrophic. The planet's oceans would evaporate, the atmosphere would become unbreathable and all life on Earth would be extinguished.

Global Warming versus an Asteroid collision. Both are approaching critical status within the same time frame. Global

Warming is creeping toward the tipping point and the Asteroid is coming. Either one is devastating.

THE END OF GLOBAL WARMING

CHAPTER 1 GLOBAL WARMING

The Beginning of a New Beginning.

2023 James, 14 years old, left home to play ice hockey at the local ice rink. Not only was it fun, but it was also cool. Ice rinks have become very popular for players, skaters, and observers. However, the cost to keep the ice rink operating continued to increase because of the increased demand and cost of electricity. Maintaining ice in hot weather is a challenge and expensive. Passing the cost on to the customers results in less customers and therefore, less income with an increasing overhead cost. The breaking point was reached, and the ice rink was closed. James and the others were disappointed. As James walked home in the heat of the day, he realized that he was a victim of Global Warming.

Global Warming is the annual increasing temperature of the surface of the Earth. I remember in the Midwest; I would be making snow angels before Thanksgiving and a white Christmas was standard. Now you're lucky to have a dusting of snow around Christmas. Why? Because the Earth is warmer. The average surface temperature of the Earth has increased 12°F degrees since the ice age 11,700 years ago and is continuing to increase.

As the Earth heats, water is vaporized creating drought for farmlands reducing our food sources and increasing forest fire probabilities. The extra moisture in the air is heated, helping Global Warming. The moisture returns to the Earth as excess rain causing storms, hurricanes, and flash flooding. As the Earth warms the water warms. The warm water contributes to the melting of glaciers raising the ocean level causing coastal flooding. Ocean life goes further north to cooler waters. Among those are sharks following their north bound food source. Result – "JAWS" become a reality of Martha's Vineyard, MA.

The warmer oceans are reducing the efficiency and life of algae. Algae produces about 70% of the oxygen in the Earth's atmosphere. Earths rising temperature is killing our best oxygen source. Algae has been reduced by 40% since 1950.

Today individuals are physically hot, seeking air conditioning everywhere. Home, to car, to work, to shopping and to home again. Rolling blackouts are occurring due to overloading of the electrical grid caused by the demand for air conditioning. Rivers and lake water levels are low. Water is being rationed for farm crops (less produce) and restricted for personal use such as watering lawns.

For the past several years the world has become aware of and concerned about Global Warming. The Earth's average temperature is being monitored along with environmental observations. A meeting was held in Paris in December of 2015 called the **Paris Climate Agreement** or **COP21**, with the goal of reducing the emission of gases that contribute to Global Warming. 144 countries attended and agreed that action must be taken to change the destructive direction the Earth is going. The objective is to reduce the rate that Global Warming is heating the Earth. The temperature of the Earth, from the pre-industrial to today, has increased 1.1°C over 140 years. The goal is to prevent increased warming of 0.4°C over the next 20 years. This may not sound like much, perhaps the difference between wearing a sweater or not wearing one on an early-spring day, but remember we are only 12°F from the ice age. The current heating rate is 0.1°C per decade. A universal agreement was signed by 195 countries on 4 Nov 2016 to limit greenhouse gas emissions to levels that would prevent global temperatures from increasing to more than 1.5°C (2.7°F) above the pre-industrial period by 2050. If this goal is not met and the global temperature soars, it could be the tipping point and life as we know it may be compromised.

This is to be accomplished by reducing greenhouse gas emissions by all countries. The prime targeted gases are Carbon Dioxide and

Methane. Fossile fuels are the primary sources of Carbon Dioxide emissions. Carbon Dioxide is about 79% of the greenhouse gases. It holds the heat and lasts a long time. Methane is about 11% of the greenhouse gases. However, it traps up to 75 times more heat in the atmosphere than carbon dioxide.

CHAPTER 2 ASTEROID DISCOVERY

NASA's Asteroid Watch program is a multi-agency to detect, track, and characterize Near-Earth Objects (NEOs). Asteroid 2023DW was discovered on February 26, 2023, by Krisztián Sárneczky at the Piszkéstető Station of the Konkoly Observatory in Hungary. The Asteroid Watch program uses a variety of telescopes and other instruments to search for NEOs. Once an NEO is found, its orbit is tracked using ground-based observatories and radar. This information is used to assess the risk of an impact and to develop plans for deflection, if necessary. They have no idea of its actual size or if it will be a hit or miss. It does look as if it will be close.

The Asteroid was discovered by ground-based telescopes. A reference picture is taken of all the stars in a location by a telescope. 4 more pictures are taken 5 minutes apart of the same area and compared to the reference picture. If they are identical there is nothing moving. However, any dot that has moved becomes a point of interest. NASA followed the chain of command and notified the authorities at Planetary Defense, a department of the National Nuclear Security Administration of the discovery.

NASA established the Planetary Defense Coordination Office (PDCO) to manage its ongoing mission of planetary defense. Under NASA Directive NPD-8740.1. The PDCO is responsible to provide timely and accurate information to the government, the media, and the public on close approaches to Earth by potentially hazardous objects (PHOs) and any potential for impact. PDCO provides information to NASA to send to the Executive Office of the President, the U.S. Congress, and other government departments and agencies.

The Asteriod is projected to possibly approach the Earth on Valentines Day 14 February 2046. The decision was not to make this information public and panic the public. After all, this Asteroid has been detected 23 years and millions of miles in advance. Planning began for what could be a direct, but survivable, hit. The destruction could be massive. "Remember the Dinosaurs!"

CHAPTER 3 SURVIVABILITY CONCERNS

There have always been some considerations of another Asteroid collision, but not taken seriously because there have been no threats. Now there is a possible threat. How do you prepare or defend from an impending Asteroid assault? Could you defend by deflecting the Asteroid with a missile or bomb? A defensive explosion could break the Asteroid into large pieces covering more of the Earth than the original. Where would you hide on Earth when you have no idea where it will hit? Homeland Security and FEMA were notified of the possibility. It was time to establish a think tank, get funding in place, and consider a contingency plan for a hit.

A classified committee meeting was held at NASA Johnson Space Center in Houston, Texas to discuss a survival plan. To understand the seriousness of the situation Dr. Samuel Dockett, Head Astrophysicist from Haleakalā 10,000 ft High Altitude Observatory, Hawaii's first astronomical research observatory, was invited to explain the reality of the situation. Dr. Dockett explained that the Asteroid could hit anywhere on Earth. As it gets closer, we will be more accurate as to where it could hit or if it will be a miss. Regardless of where it hits. "Remember the Dinosaurs." The Asteroid is 20 years away. If it hit the U.S. we would come to a halt. Power, food, transportation, and communications would be affected for years. Rebuilding would take decades. Now is time to develop a survival plan.

To develop a survival plan, the most respected personnel on the subject had to be gathered. From astrophysicists to environmentalists to construction design engineers. The committee would go forward, select key personnel, and start developing a plan.

CHAPTER 4 JAMES LIFE

News of the Asteroid discovery to a 14-year-old James would have little interest. An impact in 20 years could have been 100 or 1000 years away in his mind. Unfortunately, most of the population feels the same way. It would be someone else's problem. Live for today. Today's problem is Global Warming.

On Monday, James went to school in Pontiac, Alabama, about 10 miles from the Gulf coast. James is in the 9th grade. His father, Dan, and mother, Amy, worked at the ChipX factory just outside of town. The factory was 7 years old and produced computer chips. James' father was in charge of digital security, an ongoing challenge. His mother worked in accounting. The factory was state of the art. The clean rooms, where manufacturing and assembly are performed, were the finest in the world.

Pontiac was a small town of 15,000 until the ChipX factory came to town. Pontiac was chosen for the factory location because of the local climate and access to major Interstate 65. There were lots of clean properties. Clean meaning that the soil was very clean of contaminants that could clog air filters or be tracked in on shoes and clothing. The Gulf air was clean and fresh. The town infrastructure was good and there was lots of room for city growth and housing. The factory developed lots of green space for the town and built the ice rink. The current population is 25,000. The factory brought employment and a housing boom to Pontiac.

James and his family moved from Saint Cloud, Florida. Dan and Amy both worked at Disney. Dan worked underground maintaining computers for the above ground activities. Amy was in accounting for food management. Her job was more demanding than Dan's. The move from very hot Florida was a blessing. Especially from central Florida where there was no sea breeze.

They had a 3-bedroom house built with a swimming pool. Their first new home. All new and modern. They had lots of property. Amy made the garden she always wanted. The soil was fantastic. Her garden was very orderly. Straight rows and produce that looked like the picture on Burpee seed envelopes. Dan loved to golf and maintained his lawn as if it were a golf course.

When James started high school, he became more aware of his surroundings and what is going on the world. He ran for class president and learned about politics and favoritism. A maturing experience. He opened his eyes to events of the world. He recognized that Global Warming, he experienced all his life, had been ignored and we are now living the results.

CHAPTER 5 GLOBAL WARMING STATUS

2025 The Paris Accord Global Warming conference reported that the Earth's average temperature is still rising approaching the threshold of 1.5°C. All countries must do more to reduce greenhouse gas emissions to save the world. The goal is to be under 1.5°C by 2030. The weather is rampaging. Hurricanes, tornados, flash flooding and drought are damaging the entire food chain. Forests have turned into kindling with raging fires in Australia, Europe, Canada, Chile, and the USA. Shipping through the Panama Canal has been stopped at times because Gatun Lake water level is too low due to drought.

Aircraft are experiencing so much turbulence seat belts have been replaced by harnesses and worn all the time.

Direct Air Capture (DAC) systems capture carbon dioxide directly from the atmosphere, power plants and other industrial facilities. The captured carbon dioxide is stored underground in depleted oil and gas reservoirs, deep saline aquifers, or basalt formations. It can be used to make synthetic fuels, plastics, and concrete or to produce other products such as fertilizers, food additives, and pharmaceuticals. The use of CO_2 to synthesize products is growing. DACs are expensive to build and operate. There is no price on surviving. More and larger units are being installed throughout the world.

A next-generation nuclear power plant should be operational in a few months. It is called a Natrium reactor. It has a smaller footprint, can be refueled while operating, and has less nuclear waste. It will replace a coal plant and utilize its existing electrical grid.

CHAPTER 6 ASTEROID SHELTERS

2025 When the public was told that an Asteroid was discovered and is on a possible collision course, questions arose. How can we prepare a defense for a non-predictable impact location anywhere in the world?

The government released infrastructure funds to build new 1950 style bomb shelters. Currently there are about a dozen active bomb shelter contractors that have been building luxury shelters for years. On the announcement of the government contract hundreds of contractors appeared. These would not be luxury shelters, just functionable. There are still many 1950 style bomb shelters throughout the USA that had been repurposed for personal use but can still be used as shelters.

It takes about 2 months to build a below ground level basic 1950s type shelter. One contractor can build 6 shelters in a year. 100 contractors could build 600 shelters. In 20 years, 12,000 shelters could be built to hold about 36,000 people. The public is responsible for supplying the bomb shelters. Generators will be supplied. This appeased the public as the government is doing something that is better than doing nothing.

There are additional shelters such as underground parking garages, banks, large office building basements, subway tunnels, mines, and caves.

The government has several shelters throughout the country. One is the Raven Rock Mountain Complex (RRMC), also known as Site R near Blue Ridge Summit, Pennsylvania. Raven Rock is a massive, hollowed-out mountain. It's a free-standing city with individual three-story buildings built inside of this mountain. It has everything that a small city would have. There's a fire department, a police department, medical facilities, and dining halls. The dining facility serves four meals a day for the 24-hour facility. It was mothballed to a certain extent during the 1990s as the Cold War ended. It was restarted in a hurry after September 11, 2001, and has dramatically expanded over the last 24 years. Today it could hold as many as 5,000 people in the event of an emergency. No others are as elaborate as this. The White House, Camp David and the Pentagon have bomb shelters with communication centers.

NASA has been working on a survival plan for 2 years since the Asteroid discovery in 2023. They realized that the entire population of the world cannot be provided with safe cover. Not even in the USA. Therefore, only a select few would qualify for attempted survival. About 100 selectees would go to an underground facility, called "The Biosphere". It is designed to survive everything except a direct hit, for 3-6 months.

The master plan is to build 20 Biospheres across the USA over the next 20 years to house professionals and critical personal. Professionals would be Doctors, nurses, Biosphere design/maintenance engineers, horticulturists, and construction engineers for reconstruction/planning. The horticulturist is more important than an astrophysicist or the President of the United States (he has his own bomb shelter). All residents must be trained with unique Biosphere maintenance skills. All selectees would be DNA screened for immune diseases, celiac (compromises diets), and other diseases that are not curable because of the lack of medication, medical facilities, and no pharmacy. Contractors that have been building luxury shelters were selected for the Biosphere project to begin immediately.

Necessary for life in the Biosphere for personnel and growing food is a water source, oxygen manufacturing or scrubbing, and clean air. Three to six months plus supply of non-refrigerated food, medicines, parts for biosphere maintenance, and toilet paper. Environmental requirements are earthquake proof, endless power sources, temperature and humidity control, communications to other sites, and sanitary disposal for items that cannot be recycled.

The government realizes that only 2,000 people would be selected to survive in the underground Biospheres. Maybe 75,000 will be in shelters. That means over 300,000,000 people are unprotected in the USA. That includes the President of the United States and many high-profile personnel like Dr Samuel Dockett. Dr. Dockett accepted this as a sacrifice that he and his family would make for mankind. They could not directly contribute to the Biosphere's survival. That does not mean they would die. Where the Asteroid hits will determine who dies.

CHAPTER 7 THE BIOSPHERE SHELTER

The survival Biosphere will be built underground in three sections. Hydroponics section, a habitats section and nuclear power station. The Biosphere is designed with hydroponic foods being grown to provide fresh fruits and vegetables. The hydroponics section is temperature, humidity, and light controlled. A selection of seeds was acquired from the Svalbard Global Seed Vault in Longyearbyen, Norway. The joke is that there will be no weeds in this garden. A horticulturist will attend to the Biosphere hydroponics.

Submarine systems were chosen as the building block for the environment. The submarine gets its oxygen and drinking water from the ocean water. We won't need to. A water source was located for each Biosphere site before the start of construction. Air can be pumped and filtered by a ChipX clean air system from a tall port on the surface. A submarine nuclear reactor will provide all the electrical power for decades. The nuclear reactor will be cooled by the underground water source. Lithium batteries would provide power during reactor turbine maintenance. Everything would need to be recycled including urine if necessary. Human waste would be frozen and ejected as a solid to the surface away from the surface air inlet.

A lot of thought went into the needs for the habitat. For anyone to survive, the basic environment must provide air, water, and food. Additionally, the living quarters must be suitable for comfort and necessities. Every personal room is the same size. A couple will have the same size room as a single person. There is a need for activities suitable for individuals who will be sharing the space for a 3–6-month period or longer. The common area will be divided in the center by a LED wall. The LED wall is a display screen to display outdoor scenes, movies, or education material on either side. A tall antenna tower was constructed to communicate with other sites and to identify where the Biosphere is located below ground.

Submarine lifestyle accommodates a crew of 150. The habitat will hold 100 and be more spacious than a submarine. There will be one kitchen, 4 toilets, 2 urinals, and 4 showers. Two laundry washing machines and driers. The driers will exhaust into the air providing warm humid air. The Biosphere was never designed to be a utopia. Its existence for 3 months with a stretch to 6 months is based on food supplies. 3 months in comfort, 6 months if rationing.

CHAPTER 8 JAMES EDUCATION

2027 James graduated high school with a scholarship and started college at Massachusetts Institute of Technology to study Climatology. He wanted to learn about Global Warming that had affected him all his life. He didn't feel that he could do anything about the possible impending doomsday Asteroid collision. Rather than bury his head in the sand he would learn all he could about Global Warming because if the Asteroid passes, Global Warming will still be here. James noted that there were Global Warming naysayers and now Asteroid naysayers. And yet, the end of the world could be coming on 14 Feb 2046. 19 years away.

2029 Planning to be a survivor, James decided that he wanted to learn more about other planets and their relationships. He dropped Climatology and went to Cambridge to study Astrophysics.

At Cambridge he met the love of his life. His lab partner, Maya. Maya was born in Hawaii. Her father was Astrophysicist Dr. Samuel Dockett working on the Haleakala Volcano Panoramic Survey Telescope and Rapid Response System, the newest Asteroid tracking system.

Maya had a definite advantage over James in astrophysics, having been to the observatory often with her father.

2035 They dated and shared everything from lab to their marriage. They were married just before graduating. Both graduated with PHDs in Astrophysics. On their honeymoon they went to Maya's home in Hawaii to visit her father and for James to get some hands-on experience in his profession. He had an opportunity to observe the Asteroid that is threatening the world.

After learning to surf and climbing Diamond Head it was time to find a job. They both applied for and accepted positions at the JPL lab Arizona Radio Observatory. Kitt Peak, Arizona. Kitt Peak is a mountain in the Quinlan Mountains of Arizona. The Kitt Peak National Observatory is one of the largest astronomical observatories in the world. Although never said, Maya's father may of had some influence on their employment.

The observatory is located at an altitude of 6,877 feet above sea level. It has 25 optical and near-infrared telescopes including the Mayall 4-Meter Telescope and the McMath-Pierce Solar Telescope.

They bought their first home in Three Points, Arizona. It is about a half hour drive west to work and a half hour drive east to Tucson for shopping and city experiences. It is in the desert surrounded by sand and the homes of fellow workers. The temperature often hits 120 degrees.

CHAPTER 9 NEAR MISS

2035 The Paris Accord Global Warming conference reported that the Earth's average temperature was at the threshold of 1.5 °C above the pre-industrial period. All countries must do more to reduce greenhouse gas emissions to reverse the trend. The drought has affected growing crops and livestock. Cattle herds are reduced due to lack of feed (oats, corn, and grazing). Growers are planting further north, where the cold land of the past is now warm, with some success. but not enough. Heavy storms continue to flood and erode land. The "wet bulb" is in effect. "Wet bulb" is when it is so hot you need to sweat but you can't because the air is too humid. Results are too hot for mankind to work outside for any duration.

==

NASA and the European Space Agency Near Earth Coordination Center tracking Asteroid 2023DW had good news. The Asteroid had a Torino Score of zero. It would not hit the Earth. It would be a near miss on the outside of our orbit passing between the Earth and Mars. This message was a relief to all. Many prayers were answered.

Other good news was that the Space Based Infrared Survey Telescope overcame funding issues and was launched. It would be a year before it would be providing Asteroid detecting and tracking data. When active it will provide accurate information on how close the Asteroid will come.

Since there was no longer a threat, most bomb shelter construction, that started 10 years ago, was terminated. Many of those built would be repurposed. However, the Biospheres continued construction but at a slower pace.

Asteroid defensive guards were relaxed, and life began returning to usual. The stock market had to reorganize into what was important now. Factories that had reorganized to support the Asteroid defense projects were going back to manufacturing their home products.

Dan's company, ChipX, had converted to manufacturing medical supplies to take advantage of the clean sanitary rooms. Removing the medical equipment and reinstalling the chip manufacturing equipment is the first step to restoring ChipX. Getting everything operating to return to manufacturing computer chips again did not go well. The equipment to make computer chips is very critical. The clean room must be clean of the smallest particles with stable temperature and humidity. Meeting this requirement seemed to take forever. To get into production they had to be recertified. Certification required a production run that met or exceeded the design criteria. It took a year and two production runs before they were certified.

CHAPTER 10 NEW ASTEROID INFORMATION

2044 The Paris Accord Global Warming conference reported that the Earth's average temperature, although slowed down, was still at the threshold of 1.5°C.   2°C is critical. All countries must do more to reduce greenhouse gas emissions to save the world.   Food shortages continue to grow. Severe rainstorms, tornadoes, hurricanes, and the rising ocean are causing heavy coastal flooding. The population and manufacturing moved from the southern states north. Mosquitos and other insects migrated from Central and South America to the southern states. What is left of the power grid is strained.

==

James and Maya's had been working at the Kitt Peak Observatory for 7 years when their daughter, Emily, was born on 14 January 2044 in the hospital in Tucson.   Grandpa Dockett was on a convenient business trip to the observatory and was delighted to hold his granddaughter. There was a rumor that he may accept a position at Kitt Peak.

Four months after Emily was born James returned to work and Maya stayed home with the new family member. Interest in the Asteroid had waned since it was designated as missing Earth. James checked the progress of Asteroid 2023DW. only 2 years away. He wanted to clearly establish how close to Earth the Asteroid would pass. The tracking computer aligned the equipment to the coordinates. James was astonished to what he discovered. The Asteroid was still on course for a near miss at a million miles, but it had grown larger in size. James reported his findings to NASA Jet Propulsion Laboratory Near Earth Coordination Center.

NASA Jet Propulsion Laboratory announced the bad news from the JPL lab Arizona Radio Observatory to the public. The Asteroid, only 2 years away, was much larger than first detected and had a stronger gravitational field. The Asteroid's gravitational field accumulated items in a similar fashion to a black hole. The accumulation had increased the size and gravitational field strength since the Asteroid's discovery 20 years ago.

This was confirmed by The Space Based Infrared Survey Telescope.

> This is an M-type Asteroid. Relative to the C-type and S-type Asteroids the M-type is denser than the others. M-types are metallic asteroids normally composed of 80% iron and 20% heavy metals including a small amount of osmium, the densest element known on Earth. This asteroid did not follow the asteroid composition formula. It contained 65% osmium. The osmium made it denser. Being extra-large with a greater density created a greater gravitational field.

At first there were concerns the earth's satellites could be affected by the gravitational field as the Asteroid bypassed Earth. A much greater concern now is that the Asteroid will cause high tides with coastal flooding and will spawn high winds. Although the event is two years away, the government sees this as a threat. Now is the time to plan and take action.

Bomb shelters in preparation for a collision disaster started in 2025 had been slowed or stopped in 2035 when the Asteroid's trajectory was change from a hit to a near miss. 12,000 basic 1950s shelters had been completed. Construction of the Biospheres was slowed. 18 units had been completed and 2 others were in work at various stages of completion. The completed Biospheres were not stocked and were sealed as no collision was imminent. at that time.

Shelter construction restarted immediately to complete those stopped in 2035.   The Biospheres would be reopened, stocked and nuclear reactors were fueled.   Personnel selected for the Biospheres were notified to be prepared.

CHAPTER 11 WEDNESDAY 14 FEB 2046

No one was sure about what effects the Asteroid's gravitational field would have on the Earth. The moon's gravity has more influence on the surface of the Earth than the sun. The Asteroid is larger than the moon and four times further from the earth, but 90 times closer to Earth than the sun. Rumors varied from a beautiful view to a disaster. Warnings were to take caution, stay inside, and not stand under anything that could fall. Those with bomb shelters should stay near their shelters, including the Biospheres, just in case. Coastal areas will experience very high tides. All passenger aircraft were grounded, and the Naval fleet was at General Quarters DEFCON TWO. The President, his staff and visiting dignitaries were ordered to the White House Shelter as a precaution.

The Asteroid approached at a speed of 59,000 mph. It would probably affect the Earth for about 14 minutes. 3 minutes during approach, the first effects of its gravity may be sensed. For the next 8 minutes, while being the closest to Earth, it will be eclipsed by the shadow of the Earth. After the 8-minute period it will emerge and during the next 3 minutes, fade off into the distance. The public thought since the Asteroid was to miss by a million miles, it was going be like viewing another full eclipse of the moon that occurred 3 weeks ago on 21 January. As people stood outside watching the sky, unseen gravitational forces were beginning to cause the ground to shake. The sun was setting around 6 pm when an enormous ghost skimmed across the sky. The sun lit the Asteroid until it entered the shadow of the Earth. Some people said they felt lighter. James had warned his parents not to be outside even though there were no dangers from falling objects in their yard. Other than falling themselves, due to the shaking ground, they were not injured. Their house creaked and groaned, and the pool splashed almost dry.

CHAPTER 12 ASTEROID ATTACK

As the Asteroid neared the Earth, geo-satellites (communications, GPS, and navigation) began going offline. Those that could be tracked were followed to the Asteroid before they stopped operating. The Earth's gravitational field to hold the satellites was no match to the Asteroid's gravitational field.

As the Asteroid passed, the sun's gravitational pull tried to hang on to the Earth and the Asteroid's gravitational pull tried to steal the Earth. The Earth's surface was stretched and compressed by the two opposing forces. Two minutes into the shadow the shaking became a worldwide Earthquake. Anything with a span such as a bridge, or a dam were compromised. As the Earthquakes intensified, large hard surfaces such as roads, highways, runways cracked and buckled. Buildings began swaying. At 6 minutes the Asteroid and sun tugging on the Earth's mantle caused massive Earthquakes and volcanoes to release their energy. Dormant, extinct, and new volcanoes began weeping lava then became explosive. Hawaii's Haleakalā Observatory at 10,000 feet was destroyed when the extinct Haleakalā Volcano erupted.

Shifting tectonic plates slipped, displacing water, causing huge tsunami waves. The tsunami waves were intensified by the gravitational pull of the Asteroid. The waves splashed back and forth around the world causing damage with each cycle. The coastlines were damaged by the force of the tidal waves smashing buildings and homes of the most densely populated areas. The waves traveled inland several miles in some areas. A tsunami wave ran 10 miles up Interstate 65 reaching James's Dad's home with 2 feet of salt water.

As the Pacific Plate and North American Plate boundary (San Andreas fault) shifted, most of the southern coast of California was decimated. The San Francisco Bay area, Los Angeles, and all land points in between were destroyed. JPL labs tracking the Asteroid were destroyed in Los Angeles. A tsunami swept the remains of the southern California coast out to sea.

The world was confused about what was happening. Was this Armageddon?

On time at 8 minutes, the Asteroid emerged from the shadow leaving more damage to the entire Earth in a few minutes than all wars and past Earthquakes combined. No one bid farewell to its last 3 minutes of visibility.

Before the Asteroid got near the FAA grounded all commercial passenger carrying aircraft 36 hours before the scheduled bypass as a precaution. Non-passenger, military, and cargo express (FEDEX) aircraft would be able to fly. Air Force One would not fly because there is no enemy threat.

A FedEx flight from Los Angeles to the FedEx hub in Indianapolis was above Des Moines when it lost inflight navigation systems. GPS was lost due to the Asteroid's gravity absorbing geo-satellites. It was followed by VOR lost to surface destruction. Radio and cell phone communications were also lost. The pilot had to revert to basic pilot training navigation, dead reckoning and inertial navigation and the inertial navigation system was doing strange things. Finding a place to land was the next challenge. As he approached his destination, he could see the ground shifting and puffs of dust and dirt rise as the roads and runways cracked and buckled. Looking for another place to land he flew over Chicago, Kansas City, and St Louis. They all looked the same. With fuel running low he had to find a place to land. He was heading east over Peoria Illinois and decided that the long flat freshly plowed corn fields and no powerlines or obstacles in sight was the best invitation he would get. He made an uneventful wheel up landing near Gridley, Illinois. Many aircraft had controlled crash landings.

All train rail tracks were damaged. Rails were twisted, bent, shifted, and covered with debris. Several trains derailed. Most just stopped in their tracks. The country's train system was at a standstill.

Automobiles and trucks had road cracks to maneuver around. Bridges, overpasses and underpasses were shaken until they collasped. Streetlights and stoplights were out and the most used asset, GPS, was not working. Next problem is acquiring fuel or a charging station. Getting trucks back on repaired roads will be critical to the recovery.

Ships were in the most danger. There was no way to point the bow into the waves that were coming from all directions. All ships were encouraged to find safe harbor before the event. The tsunami created several Poseidon's. Several large passengers, cargo, oil tankers, and other ships were lost with no chance of rescue. Some small ships, yachts and small boats near shore had a small chance of surviving.

Naval Vessels at sea went to General Quarters DEFCONTWO 36 hours prior to the scheduled bypass. They would remain in DEFCON-TWO at the ships' commander's discretion. The Navy Fleet reported this was the worst sea state they had experienced. The Navy lost several ships. Ships tied up in port suffered the most damage. The fleet in Pearl Harbor was mostly destroyed.

The Asteroid gravitational field collected the Earth's moon, orbiting satellites, an abandoned space station, and some of the upper atmosphere. However, the gravitational field was not strong enough to collect the Earth, but it tried. Dr. Dockett and his colleagues, James, and Maya determined the Earth was pulled/ dragged and then released into a new orbit. The Earth was being so abused that the orbit change blended in with the violence.

CHAPTER 13 THE WHITE HOUSE

When the President emerged from the still shaking underground White House shelter, he saw devastation everywhere. Millions of people had died. All electrical power, satellites, and communications were gone. The President immediately invoked Federal Martial Law "due to a natural disaster". The challenge was how to notify the states that they were under Federal Martial Law with no communications. The White House shelter was not significantly damaged. An emergency power generator provided electrical power to operate the on-site computers and printers. Federal Martial Law declarations for each state were typed up.

The only mode of transportation was helicopters. The "1st Helicopter Squadron" at Andrews AFB, distributed the declarations by helicopter to other helicopter sites for further distribution across the nation. Within 4 days the governors and the highest military officers, Brigadier or Major Generals, in all states had copies of the orders except Hawaii. Hawaii was out of helicopter range and there was no refueling opportunities in route. The nation was under Federal Martial Law. Except Hawaii.

The president ordered all military, reserves, and the National Guard to set up security, control, and aid. Organize assistance for health care, distributing food and water and assist the local law authorities to maintain civil obedience and peace.

The radio communication center was also served by the emergency power generator. A search of the radio waves found a few ham radio operators on the air. These were the only sources of information for the White House for several days on the status of the country. In the following days communications were established with military sites and Biospheres.

The Biospheres protected and performed as designed. Due

to structural damage surrounding the remaining neighborhood most personnel elected to remain in the Biosphere for shelter.

Without satellites, communications were reliant upon the Troposphere scatter and Skywave ionosphere to communicate. Troposphere scatter frequencies are from 100 MHZ and 1 GHZ (VHF and UHF radio bands). Skywave ionosphere frequencies are from 3 to 30 MHZ (HF radio band). These bands are affected by terrain and weather conditions. Communications were spotty and weak.

Communications from surviving Naval vessels were the information source for the status around the world. The ships were to lend aid where possible. Unfortunately, the aid they could provide was limited to onboard supplies. As suspected Pearl Harbor was nearly eradicated by the volcano eruptions and tsunamis. Hawaii was notified that they were under martial law by radio.

CHAPTER 14 THE STATUS

NASA developed the survival program based on information provided by Dr. Dockett. He was their source for decisions. The climate had changed, and they needed his input. Example: breathing was more difficult, and the days are 22 hours long!

Dr. Dockett, James, Maya, and his two-year-old granddaughter, Emily, were in Kitt Peak, Arizona at the Arizona Radio Observatory when the Asteroid passed. They were hoping to get some great pictures through the telescope. The pictures did not turn out because of the shaking Earth. The lab shook until cracks appeared and equipment came loose and fell. The lab was spacious, items fell in open areas, no one was hurt. After the main shocks slowed down the three astrophysicists, using available equipment and knowledge of the stars, determined that the Earth was not where it was before the Asteroid passed. It appeared as if it was in a different orbit and further from the sun. After checking and rechecking they agreed that the Earth was about 4 million miles further away from the sun.

Dr. Dockett had notified NASA where he and his family would be located during the Asteroid event. Although there were no airplane flights there were helicopter flights. NASA sent a helicopter from Houston, Texas to Kitt Peak, Arizona, to bring the astrophysicists to NASA for advice. The view from the helicopter of the overall damage was overwhelming.

The discussion with Dr. Dockett was also overwhelming.

Dr. Dockett told them, "We suspect, by location of the stars, that Earth's orbit is about 4 million miles further from the sun. The Earth will survive because it is still in the Goldilocks Safe Zone. The Goldilocks Safe Zone for earth is 88 million to 127 million miles from the sun. Earth is now about 97 million miles from the sun.

The new orbit has its effect on the planet. We are further away from the sun. Therefore, there will be 8.2% less sunlight and heating. Solar cell output will be less. The major effect on the weather will be the temperature. The Earth is beginning to cool. As the Earth cools, rainstorms, tornados, and hurricanes will become milder. There will be very cold winters and they will last longer every year. The average temperature of the Earth may drop around 2°C over a long period of time. We will have winters like we had 100 years ago. Snow and ice. Life will go on as usual. In time the glaciers will return.

The Earths' move further from the sun increased a year to around 420 days based on 22 hours per day. That is 35 days to a month at 12 months to a year. The 22 hours day is because the rotation rate of the Earth was increased by the gravity tug-of-war between the Asteroid and the sun and the loss of the moon.

The rotation rate of a planet is initially determined by forces upon creation, such as creating the spin of a top with your fingers. Each person would start the spin at a different rate. Each planet starts at a different rate. If spinning in a vacuum without air or surface friction the top would spin forever at that rate unless it was affected by an outside force. The Earth was thought to spin with 11-hour days when created. As the Earth developed an atmosphere and water, the Earth, with the addition of the moon, slowed to the current 24-hour days and was still slowing. The moon was tidally locked to Earth and helped stabilize rotations.

This is where life becomes complicated. Rebuilding structures as well as a banking and historical system will be a difficult challenge.. All of the world's information has been digitized and put on the "Cloud". I cannot over emphasize the importance of recovering the "Cloud". It must rise from the ashes.

Recovery of the banking system is critical. Management of the banking system will be different. The old system is based on an over 1500-year-old calendar system. A new system/standard must be developed because the calendar has changed with longer years and shorter days per year. Computer systems must adapt. Clocks and watches need to be redesigned for a new 22-hour standard day or a new time standard established.

Historical information, printed materials, and guides for the restoration are needed from the "Cloud". Without GPS, we are lost. Printed road maps and travel atlases are necessary for trucks to pick up and deliver supplies from medical to materials for reconstruction.

It will take a long time to obtain and convert the old data to new standards. All historic dates, birthdays, and holidays must be redefined and adapted to the new world of time and dates. Most impacted are business records, such as banking, health records, etc. Establishing a new system of time and dates must be the most important and the most difficult from a business viewpoint.

We may still have a Global Warming problem. The Asteroid took some of the atmosphere. That's why beathing is a little more difficult. It is not serious. The air pressure has lowered to 11-12 psi. It is like beathing in the cabin of an airliner pressurized to about 7000 ft altitude. You will get used to it. Airplanes will carry lighter loads and need longer runways to take off.

The good news is that some of that atmosphere the Asteroid took contained Global Warming greenhouse gases, and the lower pressure will cool the surface of the Earth. It's too early to determine the current Global Warming status, but it should improve quickly.

This near miss was worse than a direct hit. A direct hit would have destroyed the target and an area 2500 miles around it. This Asteroid decimated the entire world.

Due to the extensive damage, our industry will have to be rebuilt.

During the rebuilding process, we need to build smart with controls on all gas emissions. It is your world, do it right or go back to the way it was just a month ago."

CHAPTER 15 ASSESSING DAMAGE

Earthquake aftershocks continued for months. The Earth resonated like a shaking bowl of jello. No one suspected this much damage would occur. Reconstruction could not begin until there was electrical power.

Reconstruction will be slow. Factories will be organized to make products for reconstruction such as building materials, wood, nails, cement, and equipment. Each factory would have its specialty. This is not a competition, but a coordination of factories   Distribution of building materials will be dependent upon trucks and railroad systems.

Although damaged, many buildings did not collapse. Structures made of wood such as homes survived because of their flexibility, but with damage. Dan's factory, ChipX, roof collapsed and cannot be used because the integrity of the clean room has been compromised and may not be recoverable.

Damage throughout the world was extensive. Many buildings and structures that didn't collapse will be demolished. Those that survived were damaged and in need of repair. Rebuilding was not just going to be a replacement. New structures would be built low to the ground, 2 story max, and as much as possible built underground.

Above ground structures would have lots of glass to collect warm sunlight and lots of solar cells. The first restoration effort for the ocean coastline will be seaports and piers.

There are 149 million single family units providing homes for 334 million people. 129 million are houses. All family units had some damage. Houses had the least damage and apartments the most damage with some collapsing into rubble.

Most house structures remained standing. The ground had shifted many times, leaving cracks in the walls, doors that don't shut right and several broken windows. Houses for the most part could still be occupied at least for shelter. Occupation is not a picnic. Utilities, water, electricity, and sewerage are a burden. Everyone's life, to varying degrees, is a disaster.

CHAPTER 16 RECONSTRUCTION PRIORITIES

POWER is the first priority. Electrical power is the base of the pyramid. From there we build.

Existing nuclear power plants auto shut down protection for the most part was successful. Because of their super construction most are repairable with minor damage. Some will take a long time to repair.

The Biospheres survived with little damage. They were well designed and constructed. Biosphere nuclear power plants are available and can support about 57,000 homes each.

We have electrical power sources from reactivated nuclear power plants, but no way to get the power where it is needed. Repair of the electrical grid for power distribution is essential. Most electric power grids, made of wires, towers, and poles, were damaged. After the water receded from the land along the coast, saltwater residue left by the ocean continued corrosion damage by attacking the metals on the power grid. Power distribution will take time to repair. Today's effort --- repairs.

We plan to build more Natrium nuclear power plants spread out in more areas. The newer Natrium Nuclear power plants can support 500,000 homes each. Using nuclear power eliminates the need to rebuild destroyed hydroelectric dams as a power source. Nuclear is a clean source of power.

Recovery will begin once the power is restored. Hospitals, homes, schools, manufacturing, and data centers will become operational. And most important, the "Cloud" will begin to rise from the dust. Libraries of information will become available.

Important for the reconstruction is printing material and reference guides, and road maps for trucks and trains until GPS satellites are available. Trucks and trains will become our best friends delivering food, medical supplies, tools, equipment, fuel, water, clothing, etc. Repair of roads and bridges for trucks and trains is a priority. Rebuilding the highways will look like WPA or CCC projects of the 1920s and 1930s. Mostly by manpower.

Electric power will start the petroleum refineries to produce fuel and oil for trucks, trains, building equipment, manufacturing, heating oil, and farm equipment. Existing fuel will need to be rationed until it is in full production. It's to be used for essential, emergency, and farm equipment.

Require communications/internet with or without satellites. Rockets launching satellites is not in the near future. Repairing telephone towers and recovering telephone systems as soon as possible is necessary.

Community water utilities are needed.

Global Warming forced farming and livestock to migrate to the northern states and Canada. It won't be long before it's too cold to grow crops in the northern states. The farmlands in the southern states were abandoned and have gone through droughts and flooding over and over and are badly eroded. Restoring the farmlands is a priority. Food must be planted, grown, and harvested as soon as possible. Beef cattle, hogs, and chickens must be raised. Until power for construction and manufacturing are functioning the farmer needs hands-on manpower help.

Many hospitals, although badly damaged, had backup generators that still worked. Some generators and equipment were dug up from collapsed hospitals, banks, and other sources. Recovered power generators were distributed to areas of significant importance such as gas stations. Gasoline / diesel fuel was only available to first responders, construction equipment, repair units and farm equipment. Reconstruction priorities are hospitals, schools, and housing.

The reality is that we are 2 to 3 years away from electricity and fuel distribution. Until then we must provide food and shelter.

CHAPTER 17 THE BEGINNING OF A NEW BEGINNING

2058 Over a period of 12 years the new world is beginning to recover. Recovery will take generations. The government continued under Martial Law until 2055 when rescinded. The presidential election is scheduled for next year.

The remaining world is alive and wanted to continue living. People worked to rebuild and improve their lives and others. They enjoyed the new cooler weather. Many young people had never seen snow. Food supplies became abundant as the dried-up rivers and lakes were filled with water again. Dried up farmlands and trees of 12 years ago were flourishing. The cooling Earth is in balance providing gentle rains and drought is in the past. However, the sun is a little weak.

Cities began to take shape. Hospitals, schools, and manufacturing facilities were being built. Data centers were becoming operational.

Offices were assembled for management of reconstruction. Modern building technology expedited construction of living quarters and offices. Apartment housing was a priority. Offices and apartments were assembled with modular units like cruise ship's modular units. Two paired units per living quarters. One unit is living space and the other is the kitchen/bathroom. All kitchen/bathrooms are connected to a common plumbing and electrical power distribution area. The kitchen/bathroom units can be connected or removed as required. The living space interior designs are flexible. The units are pre-built to selective or custom design within strict combability parameters. Each unit is air-conditioned/heated and sleeps 2 or 4 based on your needs. The new owners must furnish the interior. A two-story building of 20 units could be assembled in 60 days. The modules were manufactured by several companies to meet the housing demand.

Most single homes were repairable. James's parents' home had minor damage and was salvaged. James and Maya and their daughter, Emily, and Grandpa Dockett moved in with James's parents when the house was livable again. Since the house had only three bedrooms a modular bedroom was added. Emily, now 14 years old, enjoys the repaired swimming pool with her friends and her own bedroom. With Global Warming under control, she is experiencing real seasons and cold weather. Something James never had during his Florida childhood.

James and Maya became Professors in the new University teaching what they know best. Astrophysics.

Dan's factory, ChipX, was restored and operational this year. The local population of Pontiac assisted with rebuilding as much as possible. They had a vested interest in future employment. Amy's garden took on the perfect Burpee seed envelope look again. Amy and other gardening women set up a farmers' market on weekends. This was the new norm.

The completed 20 Biosphere habitats units are being used as temporary electrical power sources and are reserved for emergency disasters.

Global Warming measurement had dropped to the 1.1 °C level. That is the same level as it was in the year 2000. Several things helped this happen. Although the Asteroid was responsible for the destruction and loss of millions of lives, it also destroyed factories and other sources generating greenhouse gases. Managing emission gases was taken seriously by reducing carbon dioxide and methane during reconstruction. Construction equipment used catalytic converters to clean their exhaust. Factories use scrubbers to remove pollutants and other harmful substances from the exhaust gas before being released into the environment. There were also efforts, throughout the world, to lower the level to 1.0°C.

Repaired Direct Air Capture (DAC) units continue to capture carbon dioxide directly from the atmosphere and from power plants and other industrial facilities. The few remaining coal powered generating plants were replaced with clean Natrium nuclear power plants. Carbon dioxide emissions were reduced by the using alternative fuels, such as electric vehicles, hybrid vehicles, and vehicles that run on natural gas or hydrogen. A big help was that only a few people were driving because there were only limited roads and most worked near their homes.

The Asteroid moving the Earth over 4 million miles farther from the sun reduced the sunlight by 8.2%. With less sunlight and lower atmospheric pressure, the Earth began to cool. The cooling Earth cooled the oceans. Algae in the cooler oceans is recovering and absorbing more greenhouse gases, especially carbon dioxide and producing more oxygen. There is a future.

The Asteroid moving the Earth further from the sun ended Global Warming. Mankind on its own was not able to manage their created Global Warming disaster. It took an outside force to change the course to extinction. Global Warming poses an existential threat to the long-term sustainability of human civilization and the natural world upon which it depends. Mankind has been given a second chance.

The END of Global Warming.

EPILOGUE

It's ironic that the Earth's temperature can be controlled by Global Warming. Global Warming could be the thermostat to prevent the next Ice Age.

Raymond Boyd

He was born in Pontiac Illinois, a small farming town. He joined the Navy at age 17. He was trained on Naval Weapons Fire Contol Systems. His first military assignment was on the development of the Submarine Fleet Ballistic Missile System at Cape Canaveral. After the Navy he worked in industry supporting military aircraft attack weapon systems and Electronic Warfare Weapon Systems for 42 years. He supported the military aboard aircraft carriers and in Vietnam.

I've Had The Lifetime Of My Life

This is a memoir of my life from childhood to retirement. My experience in the Navy during development of the submarine Polaris Missile system and experiences after the Navy from a shipyard worker to supporting military operations in Vietnam, Spain, CONUS, and aboard aircraft carriers.